House Beautiful

WALLS & FLOORS WORKSHOP

House Beautiful

WALLS
& FLOORS
WORKSHOP

TESSA EVELEGH

HEARST BOOKS
A Division of Sterling Publishing Co., Inc.
New York

Created, edited, and designed by
Duncan Baird Publishers Ltd, Castle House,
75–76 Wells Street, London W1T 3QH

Managing editor: Emma Callery
Designer: Alison Shackleton

Photographs: see credits on page 157. The publisher has made
every effort to properly credit the photographers whose work
appears in this book. Please let us know if an error has been made,
and we will make any necessary changes in subsequent printings.

Library of Congress Cataloging-in-Publication Data
Evelegh, Tess.
House beautiful walls & floors workshop / Tessa Evelegh. p. cm.
Includes index.
ISBN 1-58816-406-3
1. Interior walls--Decoration. 2. Floors. 3. Floor coverings.
I. Title: Walls & floors workshop. II. Title: House beautiful walls and
floors workshop. III. Titles: Walls and floors workshop. IV. House
beautiful. V. Title
NK2119.E94 2005
747'.3--dc22   2004054041

1 2 3 4 5 6 7 8 9 10

Published by Hearst Books
A Division of Sterling Publishing Co., Inc.
387 Park Avenue South, New York, NY 10016

House Beautiful is a trademark owned by Hearst Magazines
Property, Inc., in USA, and Hearst Communications, Inc., in Canada.
Hearst Books is a trademark owned by Hearst Communications, Inc.

www.housebeautiful.com

Distributed in Canada by Sterling Publishing
c/o Canadian Manda Group, 165 Dufferin Street
Toronto, Ontario, Canada M6K 3H6

Distributed in Australia by Capricorn Link
(Australia) Pty. Ltd.
P. O. Box 704, Windsor, NSW 2756 Australia

Printed in China
ISBN 1–58816-406-3

CONTENTS

FOREWORD

Planning a new look for any interior brings with it that wonderful sense of anticipation. A fresh new look can influence not only the appearance of a room, but how you use it. At this beginning stage of interior design, walls and floors take priority—bridging the gap between the architecture and all the moveable elements such as furniture and furnishings. Even though walls and floors are static and you're changing only their outward look, how you do so will have a huge influence on how you use the space.

The carpeted floor that was essential for babies and toddlers may be better exchanged for vinyl as the children grow up—to make for quick mopping up of spilt drinks. Or, a large open space may need to be rethought—zoned, perhaps, into more functional kitchen and eating areas. This can be readily done simply by using different floor surfaces. Put stone in the kitchen area where you dine, for example, and you'll begin to see and use the space differently. Wall color too can be used to influence the appearance of space. With the right choice of color you can make the room appear more intimate or more open and airy; wider or narrower, the ceilings can appear higher or lower.

In addition to affecting the architectural look and use of space, walls and floors offer plenty of creative scope. Even if you feel your artistic skills begin and end with wielding a paint-filled roller, you'll still have plenty of room for expression. So spend time discovering your favorite color combinations; decide whether you prefer every surface painted in the same shade, or whether you want to make a statement by painting a single wall in a contrasting color. Consider decorating your wall with motifs—it's easy using ready-made stamps. Or create unique geometric designs using masking tape. If you're more ambitious, try special effects, stenciling, or even mural painting the walls. If you

don't want to tackle this yourself, you can find mural artists to do the work for you. The trick is knowing what you want and showing drawings or photographs of the effect you are trying to create to the artist.

Floors also offer plenty of scope for creativity, even if that simply comes down to making choices: hard or soft floor; timber, stone, or ceramic; and the size and arrangements of tiles. The right color is also important—whatever the floor.

This book will help you address all of these issues. Its pages are filled with hundreds of photos of beautiful rooms to inspire you. You will also find the information you need to help you ask the right questions in the showroom. With *Walls and Floors* you can be sure that you will make the right choices for you and your home, and that the results will be both beautiful and liveable.

MARK MAYFIELD
Editor in Chief, *House Beautiful*

FIRST DECISIONS

FINDING YOUR STYLE

Our homes are a natural expression of ourselves. Over the years, we add to our possessions and build up collections. We address our changing lives by developing and adding depth to our style. As fashions change, we adapt our homes, adding color accents, textures, or patterns to update them. The key, at the outset, is to set a basic style that you'll be happy with and that can be developed over the years. Try to give yourself time: in the haste to get the house just right we often make decisions we later regret. Look through books and magazines to help you decide what you like, and make up a file or storyboard of your favorite pictures. You'll soon see a theme emerging on which you can base your overall look, so that you can begin to plan the walls and floors.

▲ Classic elegance
Pretty curvy lines and discreet detailing are the hallmarks of the traditional Swedish style that gave rise to the delightful New England look. Floors and walls are kept to neutrals, whites, gray-blues, and greens.

▶ Sleek modern
Clutter-free surfaces and low sleek lines are the key to modern style. Walls and floors are usually plain rather than patterned and aim at enhancing the architecture of the building.

FINDING YOUR STYLE: flooring

Installing a new floor is not something you'll want to do regularly: few are cheap and laying them is disruptive as the whole room has to be cleared before the work can begin. For this reason, choosing new flooring is a major decision, especially if you plan to lay a hard floor, such as wood, stone, or ceramic tiles. Damaged original floors that are an intrinsic part of the architecture should be restored if at all possible. If not, try to reinstate one of a similar style.

If you want to put down a new hard floor, try to make a classic choice that won't date and will comple- ment wall coverings of different styles. Very light, very dark, and highly patterned floors can make wonderful fashion statements but are probably not the best investment in the long term.

Soft and resilient floors are easier and more economical to change on a regular basis, and can be expected to last from five to 20 years. If you want to use the latest vibrant colors or patterns, choose inexpensive vinyls and carpets, which can be changed at your whim as interior fashions move on. Details of each of these floorings are given on pages 104–151.

▶▶ **Antique beauty**
Nobody would advise you to put a wooden floor in a bathroom, especially if the wood contains large cracks. But the beauty of this ancient oak floor has been acquired over time, and could never be reproduced with new materials. Changing or covering it would only ruin the character of the room.

▶ **Modern moves**
These polished stone tiles have an undeniably contemporary feel. The choice of an elegant neutral (in this case, a greenish-gray), means they should withstand the test of time.

FINDING YOUR STYLE: walls

Do you want the walls to be a background "canvas" that can be used to set off the rest of the interior scheme? Or would you prefer to make a statement of the walls? Plain colors are the best way to make a backdrop, and while neutrals are popular, they aren't your only choice. By painting a room in a single color, including ceilings and woodwork, you can create a harmonious setting for furnishings and fabrics. This was a favorite Georgian style, with even the closet doors painted to match the rest of the walls.

If you want to go to the opposite extreme and use color or pattern on the walls to make an interior statement, you'll need to be more restrained in your choice of furnishings. Or you can paint a "focus" wall, picked out in contrasting color to the rest of the walls to highlight a special piece of furniture or sculpture.

◀ Background style

By painting everything on the wall in one color, you can create an harmonious whole, against which the rest of the furnishings can then be set.

◀◀ Making a feature

A single wallpapered wall makes a focal point in this elegant Swedish style room. All the other walls, the ceiling, and the floor, are in white, making the blue wall even more effective.

USING COLOR

More than any other element in the room, it is the color that sets the style (see previous page). Use it cleverly and you can make a difference to the look and the ambience of a room. There are two ways to approach color: you can decide at the outset what colors you want your walls to be, or you can be guided by favorite furnishings. Sometimes, it is easier to do the latter, matching walls to furniture because there is virtually no limit to the paint colors available, and it should be easy to make a computer match. However, upholstery is restricted to whatever shades and patterns are on the market at that moment, so you may not be able to find a fabric that works well with your chosen wall color.

Start by researching what looks you really like (see page 10). Neutrals are elegant, but if you like brighter colors, think about how you'd like to use them. Do you like the aged look of paint effects (see pages 70–71), or do you like matte color? Do you like solid color throughout the room, or do you prefer to paint different walls in different colors? Strong colors can look fresh and very striking if used on only one wall or you can use two harmonizing shades in a similar manner.

◀ **Neutral peace**
Whites and creams are perennially appealing because they evoke a sense of peace and harmony. Furthermore, by choosing furniture and furnishings of the same hue, you can successfully mix and match quite disparate styles. This also makes it easier to adapt to changing fashions.

▶ **Bright style**
Strong colors can be invigorating. For a fresh, modern look, paint a single wall in an otherwise white room in a bright color. This is easy to live with, sharpens the effect of the color and, furthermore, can be used very effectively to visually change the proportions of the room (see page 22).

PATTERN

The taste for pattern in interiors changes with the decades, but the principles of pattern remain the same. Pattern is used to break up the monotony of plain schemes, often bringing flair to the interior. It can be discreet, using tiny motifs, almost as a texture; or it can be strong, designed to make a statement. Stripes have always been smart: fine and subtle, or bold Regency style. Florals usually suggest a country feel, though they can be teamed with stripes for a sleeker look. Big, bold patterns are the reserve of the brave, but, carefully chosen, they can transform a tasteful, plain interior into one that undeniably demonstrates flair. Pattern is usually happier on the walls than on the floor, as highly decorated carpets can become distracting. Patterned rugs, on the other hand, whether they're traditional eastern style, or modern state of the art, can be one of the most successful ways to add interest to a classic stone or wooden floor.

▶ **Floor Flair**
A bold 1920s carpet adds flair to a sophisticated color scheme, unifying a variety of contrasting motifs.

▼ **Stripe it right**
Stripes never look fussy but they can be used for many different finished effects. Here, fine blue stripes look fresh and pretty set against white in the bedroom.

TEXTURE

Texture provides an alternative to pattern: perfect for those with plainer tastes who want to add interest to a room. One way to start is with the floor covering. Texture is a distinguishing feature of many carpets, ranging from sumptuous shag piles to knobbly, bobbly-loop piles and luxurious velvet. Natural floor coverings, too, have intrinsic texture. The fibers are ruggedly textured, and this is accentuated by interesting weaves, such as herringbone, basket, and twill (see page 144).

Rubber flooring can add texture in quite a different way. You can choose from a wide variety of surfaces ranging from shiny and flat to studs and ridges in various sizes (see page 136).

By making use of smooth, glossy finishes, you can provide valuable contrast to deeper textures. Here again there are plenty to choose from: highly glazed tiles, polished marble, granite, and limestone, stainless steel wall panels, mirror, and glass.

▶ **Tactile walls**
White painted bricks have a rugged appeal on the original walls of this old house. The texture is accentuated by the smooth, shiny, gilt-framed mirror.

▶▶ **Natural texture**
Texture can be introduced with found objects. Here, mussel shells, collected over one season, have been cemented onto a chimney to form a loose mosaic, bringing natural pattern and texture to a simple room.

PROPORTION

Visual tricks can seem to alter the proportions of rooms and there is no better place to start than with on your walls or floors. The rule of thumb is that lighter colors tend to visually push walls outward, and darker or stronger ones bring them in. So if, for example, you have a long, compact room, you may want to paint one of the narrower walls in a darker or brighter shade than the rest of the walls, thereby bringing it inward. Another trick would be to hang a large mirror on one of the long walls to reflect the light and create the illusion of added width.

Ceilings, too, can have an effect on the proportions of the room. A high ceiling in a small room can make the walls look overbearing, so aim to bring it downward visually by painting it in a slightly darker shade than the rest of the walls. Also, avoid ceiling lights. Instead, fit wall lights as these will visually push the walls outward while bringing the eyeline down, away from the high ceiling. A low ceiling, on the other hand, which makes the whole room feel oppressive, can be given a lighter and more airy feel with visual height. Do this by painting it a lighter shade than the walls.

Floors can also be used to visually change the proportions of a room. Generally, the paler the floor, the more light is reflected upward, making the whole room appear larger and more airy. By contrast, darker floors can be used to create a cozier look. For long thin rooms, use two rugs, each "grounding" its own furniture to make two pleasing spaces, rather than have one that is less than appealing. Where function is the determining factor, as in a combination kitchen and dining room, consider using two different materials—for example, stone in the cooking area and wood in the dining.

▶ **Doubled up**

Use mirrors to reflect and double the natural light levels in a room. These full-length mirrors each reflect the light from windows on either side of the room, creating a bright and airy atmosphere.

WORKING WITH LIGHT

The way in which natural light reflects off wall and floor surfaces can affect both the ambience and the perceived proportions of a room. To make the most of the natural light within a room, paint the wall opposite the windows in white, or hang a large mirror opposite the natural light source.

Another trick is to channel the light from one space into another. The most obvious dark areas in many homes are halls or corridors where there is little direct light. One solution is to fit glass windows, doors, or parts of walls between the hall and adjacent rooms that are filled with natural light.

If privacy is a priority—for example, for bathrooms—put windows high up so the light can still travel through to the corridor but nobody can look in. Glass doors and windows are often fitted between rooms so that the light can travel. A modern alternative is to replace sections of the wall with glass bricks. They are more substantial than windows, yet let in plenty of natural light.

▶ **Travelling light**
By making a passage through either side of the fireplace, light is allowed to flood from one room into another, substantially increasing the light in both spaces. The pale wooden floor enhances this and at the same time unifies the different areas.

▶▲ **Borrowed light**
Light can be directed from one floor onto another. This lovely gallery lets light travel from two downstairs windows up onto a dimly lit landing. The whole area is painted with a pale lime-style varnish to reflect the available light.

▶▶ **Light times**
The morning sun makes this an ideal breakfast room. Rather somber at other times, with the arrival of day it is most welcoming.

ROOM
BY
ROOM

LIVING SPACES

In traditional homes, living space comprises at least two rooms: a family sitting room and a dining room. Some homes also have a second, more formal living room, reserved for entertaining guests. Contemporary homes are more likely to have a single open plan space or great room that can be used flexibly. Seating and dining areas may occupy opposite ends of one room, which may also incorporate a workstation. Such spaces with multiple uses require well thought out flooring and decorating solutions, both to "zone" the areas and to suit their various ends. In the living area, for example, the priority is comfort, with soft flooring and relaxed decoration; while in dining and work zones, there may be more emphasis on practical, easy-to-clean surfaces.

▶ Harmonious color

Many Victorian houses have two living rooms downstairs. Nowadays they are often opened up into one large space. Here, one end of the space is used for dining; the other for sitting, and the whole space has been given a harmonious feel by decorating throughout with aqua.

▼ Seating times two

This living space is divided into two seating areas: informal, for television viewing, and formal. The space is united by the same polished wood floor, but each zone has a separate floor rug. The division is further emphasized by a "floating" wall in café au lait, echoing the cozier television area, in contrast to the white formal area.

PUTTING IT TOGETHER:
dining spaces

Dining spaces that are used mainly in the evenings demand an intimate feel. Deeper colors work well because, when the lights are low, they appear to fall away into darkness, adding to the intimacy. Taking the focus away from the walls also lets you be more adventurous with the other colors. Dining room floors can be covered with almost anything. They don't have to be soft underfoot like a bedroom, non-slip like stairs, water-resistant like bathrooms or kitchens, or even able to withstand hard wear like halls, stairs, and landings. So the choice is highly personal—anything from original flagstones to shag pile carpets would be appropriate.

▲ Clean simplicity
Flexible dining areas that serve other purposes during the day work best within a restrained scheme. Here, the simplicity of the restored oak floor and white walls creates a subtle background that makes a busy room easy to live in.

Cream architraves and window frames are an elegant choice set against deep-colored walls. Paired with the café noir they appear fresh and restful—pure white would have looked stark.

Café noir walls are a sophisticated choice for a dining area, creating an intimate ambience for elegant evenings.

The buttery tinge of the doors brings light relief to the walls, echoing the color of the dining room furniture.

The dado rail is painted the same color as the rest of the wall for a sleeker finish. If it had been picked out to match the architraves, the room would have been visually cut in half, and the overall look far busier.

Here, cream paint is applied in a checkered design over wood to introduce pattern to the floor. The cream squares go well with the cream woodwork, while the natural timber refers to the original architecture.

The dramatic medallion, depicting a golden rooster against a deep chocolate background, echoes the wall colors. And it makes a striking focal point.

▲ Coffee and cream

Coffee colors are very sophisticated, and can be lightened, as done here, with creamy shades. The paler floor and furniture reflect the light so that there is a soft glow to the walls. The beige checks complement the cream woodwork.

KITCHENS

Easy-clean surfaces are the priority in kitchens. Although the 1970s saw the introduction of kitchen carpets, it was hardly a serious proposition. Deep pile has never been compatible with spilled food. Instead, stone and ceramic are the traditional favorites (see pages 112 and 116), and they have now been joined by wood, which can be finished with easy-clean varnish (see page 120). Resilient floors are also a popular choice as they are generally softer underfoot and easier on the budget (see pages 129–137).

Kitchen walls are different from others in the house because a good number of them are covered by cupboards or shelves. You can go one of two ways—choose colored cupboard doors and paint—or tile—plaster surfaces in a white or neutral shade. Or you can choose neutral units, reserving color for the areas beneath. The second option offers more flexibility as interior fashions change.

▶ **Plan for change**
These flat-fronted, unembellished beech cabinets are a perfect choice, as they are unlikely to date. It is the red-painted walls that provide the style, which can be changed relatively cheaply and easily when fashions change.

▼ **Wooden heart**
Black and white checkers add interest to the original floorboards of this kitchen, tying in nicely with the black and white furniture. The tongue-and-groove walls have been painted white for a clean, uncluttered look.

◀▼ **Classic choice**
White walls teamed with wooden cupboards are a kitchen classic. It is on the floor of this kitchen that a fashion element has been introduced. The original floorboards have been painted blue with a white coachline border and finished with a protective varnish.

PUTTING IT TOGETHER:
kitchens

When planning kitchens, the best solution is to choose a classic, simple design for those elements that are either expensive or difficult to change, such as the floor and kitchen cupboards. In that way, your kitchen is less likely to date. You can then introduce the fashion element with paint on the walls or less expensive resilient flooring. Floor and wall cupboards are inclined to divide the kitchen horizontally, so counteract the effect with some vertical elements, where at least part of the wall is a single color from floor to ceiling.

▲ Color link

A lime-green varnish that matches the walls was used on the floors, creating a sleek uniform look. White was chosen for the kitchen cabinets, linking them visually with the furniture. White is a classic choice for cupboards as it teams well with any future color changes on walls or floors.

◄ White winner

Clean white surfaces imply good hygiene, but there's a reluctance to use white near cooking areas for fear of grease marks and splashes. This white kitchen circumvents the problem by using wipe-clean surfaces in the form of glazed ceramics and eggshell paint.

White tiles carry through the monochromatic theme.

All the wall surfaces and cupboards are painted white for a crisp, cohesive finish.

Glazed ceramic tiles make a durable kitchen floor.

Large floor tiles set diagonally give both the illusion of space and a dynamic quality to the whole room.

BATHROOMS

Water-resistant surfaces are priority in the bathroom—floors and splashbacks that won't be subject to leakages, as well as walls that can be wiped clean of water splashes. This accounts for the age-old popularity of ceramic tiles, which do both. The Romans used them extensively, and they are a favorite of modern architects, too. Another reason for their appeal is that they are available in a huge variety of styles, from tiny mosaics to large slabs, in shiny to matte glazes, and in colors that span the rainbow (see pages 93–101).

Recently, colored glass slabs have provided an alternative to wall tiles, but they are still available only in specialty stores. Stone is another favorite choice for both bathroom floors and walls (see page 112). Resilient floors and wipe-clean vinyl wallpapers designed specially for bathrooms are gaining popularity, although they are less permanent.

▲ On the tiles

Walls and floors can be covered in the same tile design, though the floor variety comes in a heavier weight than the wall. Here, tiny mauve tiles used throughout the bathroom look both modern and architectural.

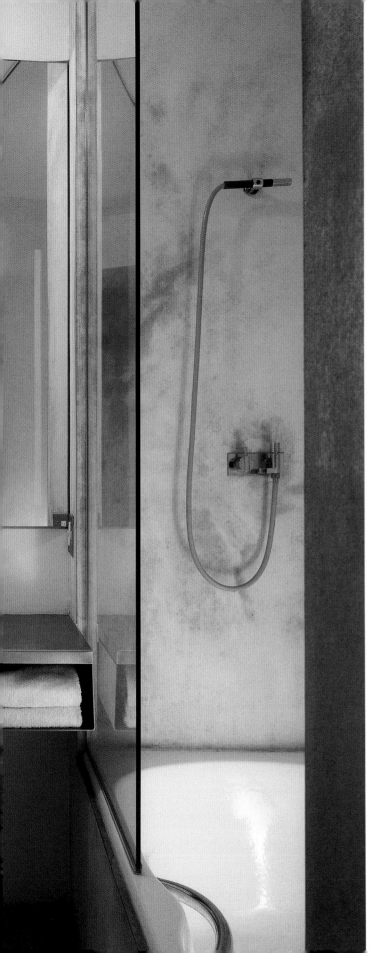

▲ **Team spirit**

In this handsome bathroom, some quite diverse materials are linked by white. A marble tiled floor, glazed ceramic wall tiles, and latex paint make a refreshing combination.

◀ **Size matters**

Tiles of different sizes can be used to good effect in a bathroom. Mosaic tiles create an all-over geometric pattern on the walls by virtue of the grout; larger tiles, like those used here, look sleek and elegant.

PUTTING IT TOGETHER: bathrooms

White is an ever-popular bathroom choice as it teams well with classic white porcelain sinks and commodes. Ever since colored sinks lost favor in the 1980s, most bathroom designs have incorporated some element of white. By using different materials that reflect the light in different ways, it is possible to create a myriad of different looks using just white alone. However, white does not have the monopoly on bathroom color schemes, and the increasingly large range of wonderful colored tiles offers endless scope. By putting colored tiles on just one wall of an otherwise white bathroom, for example, you can create a striking, modern look.

Walls are painted white for a striking contrast to the navy tiles.

White cupboards match the white painted wall for a cohesive look.

▲ Sheen and shine

The shiny surface of a ceramic tiled floor reflects light into the room. Set against the elegant sheen of white eggshell painted walls, it makes a sophisticated combination.

Glass bricks are a waterproof alternative to tiles, allowing light to travel from one place to another for an overall brighter space.

Large navy blue ceramic tiles create a sleek bathroom floor.

The navy tiles continue onto the bath and wall surfaces for greater cohesion.

▲ White alternative

Mosaics are attractive as bathroom floors. This one teams well with white ceramic wall tiles laid in classic brick bond style.

◄ Chic contrast

Navy and white is a handsome contrast that is always successful. This crisp contemporary solution works especially well with the introduction of glass bricks to increase the light levels.

BEDROOMS

Comfort and relaxation are the most important elements in decorating a bedroom. Most people want to step out of bed onto something soft underfoot, be that a wall-to-wall carpet or deep pile rug. Since bedroom floors aren't subject to heavy wear and generally do not open onto the garden, they're not subject to as much soiling as other rooms. This means that wall-to-wall and fully fitted carpets will work well. When it comes to walls, the most successful schemes are restful and harmonious, aimed at relaxation and inducing sleep. Paint is the least distracting wall finish, but wallpaper is also popular because bedrooms can get away with being ultrafeminine: a look that can only be enhanced by pattern. Strong contrasts are best avoided as they are more inclined to be stimulating rather than restful.

▶ White simplicity

The elegant architecture of this old house is beautifully shown off with pure white paintwork on the walls and careful restoration of the original oak beams. The soft white carpet makes an inviting surface to step onto as you get out of bed.

▼ Sumptuous comfort

Wall-to-wall carpets provide the perfect landing for bare feet, and create a seamless, restful look for the whole room. This neutral color scheme adds sophistication to the whole setting.

PUTTING IT TOGETHER:
bedrooms

Walking barefoot on something sumptuous is a wonderfully sensuous everyday experience—but that does not mean we all have to have wall-to-wall carpets in the bedrooms. If you have a beautiful original oak floor, for example, you may not want to cover it up. This is where a gorgeous deep pile rug on either side of the bed may come into play. It's a matter of assessing what you have and making it work for your home. Walls can be kept plain and simple with paint, or given a pretty, feminine effect with sprigged wallpaper. Fabric wall coverings are another option. They give a supremely luxurious feeling, which is perfect for adult bedrooms, but should be kept out of rooms where little sticky fingers could cause harm.

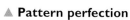 **Pattern perfection**
Floral patterns are charming when teamed with geometrics. In this bedroom, the simple striped rug perfectly offsets the ultrafeminine wallpaper.

Layers of polish applied to the original wooden floor over the years have built up a rich patina, adding to this room's sense of history.

Understated patterned wallpaper introduces a sense of intimacy and interest to the walls.

A simple striped rug brings softness to the floor, adding a contemporary accent.

▲ Sweet simplicity

A simple rug and a lick of paint is all it took to turn a rustic room into a bedroom of immense charm. A beautifully harmonious setting was achieved by painting everything, beams and all, in neutral shades.

◄ Classic comfort

The combination of softness underfoot and the subtle patterns on the walls makes the decor of this bedroom both timeless and relaxing.

CONNECTING SPACES

Halls, landings, and stairs are areas of high traffic that everyone passes through to get to other rooms, so they take a lot of wear and tear. Halls also take the lion's share of dirt from the street, but you can counteract this with a generous piece of coir matting, preferably fitted into a sunken area to keep the matting in place. All connecting areas need hard-wearing floor coverings. It's always better, if at all possible, to refurbish original stone, ceramic, and wooden floors, rather than cover them with a less resilient flooring. Wooden stairways, on the other hand, take the most beating of any areas in the house and would soon become worn and damaged if not given some kind of protection, be it a runner or fully fitted carpet.

 **Long-term solution**
Original flagstones have served this farmhouse well for more than a century and still provide the best flooring solution. Here, they are complemented by exquisite cream painted walls.

▶▲ Light moves
White ceramic tiles make a handsome, light-reflecting yet resilient hall floor. At the same time, the sunny yellow walls help to reflect the light through the stairwell for a warm, welcoming entrance.

▶▶ Classic inspiration
Instead of traditional ceramic tiles, this black and white checkered floor is made from high quality, easy-to-lay vinyl. The hall is painted white throughout for a clean, airy result.

PUTTING IT TOGETHER: connecting spaces

Hallways are often starved of light as they are usually in the innermost part of the house, so a priority for inner walls is to keep them bright. You can do this by using light or vibrant colors or by hanging mirrors to reflect any available light. Crisp white is always a successful hall color, as are all the light, bright shades of yellow. Green, too, is often used, perhaps reflecting a subconscious desire to bring the outdoors in. For the same reason, it is also a good idea to make floors as light a color as is practical, using shiny finishes, such as glazed ceramics and full gloss varnish, which help to reflect light throughout the space.

▶ **Light fantastic**

The use of modern materials and pale neutral colors make this landing floor light and bright enough for it to double as living space.

▼ **Travelling light**

White walls and cream carpet on the lower floor reflect all the available light. In this case, the light is plentiful, generously flooding down from the galleried upper level.

White paint throughout reflects the natural light and maximizes it.

Translucent screens allow the light to be shared between the two spaces.

Solid oak finished with industrial quality varnish provides a hard-wearing and easy-to-clean floor covering.

WALLS

INTRODUCTION

Walls play the major role in the interior, exerting greater influence on the overall look of the room than any other element. They provide a background "canvas" that can be either restrained and aimed at setting off the furnishings; or designed to make a bold statement using color or pattern. Walls can also be used as a fashion statement because they are readily adaptable and changeable. While buying new furniture is usually seen as a long-term investment, depending on the state of the walls, they can be given a whole new look with an application of paint over the course of a weekend. You don't even need to paint all the walls—a statement color on one wall in a white room, for example, can be changed with the seasons. Walls can also be given a unique tailor-made finish using a special technique. Although the application is painstaking, the end result is unbeatable in terms of personal expression.

▲ Low-tech statement
The building finish on walls can be used to make a designer statement. This robust look is achieved by polishing the basic plasterwork.

◀ Dramatic paint
Paint offers the best value for money and the quickest way to change the look of a room. Furthermore, it doesn't have to be boring: it can be applied in blocks, stripes, or outsized geometric shapes. Although this is a painted canvas, blocks of color painted directly onto the wall will have the same dramatic effect.

◀◀ Pretty impressive
Wallpaper quickly adds pattern to a wall. Pin up a small swatch in a room before buying and you will immediately know what the final outcome will look like.

PAINT

color

texture

quick

easy

variety

creative

bright

subtle

moody

cool

muted

QUICK COLOR

The joy of using paint is that, given well-prepared walls, you can quickly transform a room. Not only is there a vast array of colors to choose from—which change with interior fashion—but there are a myriad ways in which to apply them. You can use a single color throughout the room; pick out architectural features in white; or paint different colors on different walls. Flat color may be used one year, followed by silk finishes a few years later. For the high fashion end of the market, manufacturers have introduced specialty paints with metallic or pearlized finishes.

Although most stores carry a limited range of ready-mixed paints, thousands of shades can be mixed to order. Some stores even have a computer sensor system that can analyze the color makeup of a swatch of fabric, for example, and mix paint to match.

▲ **Modern solution**
Strong, flat colors introduce modern appeal to even the simplest of rooms. Choosing eggshell rather than a gloss finish for woodwork and shelves produces a clean and unified effect.

Delicate celandon green latex on the upper walls is given decorative detail with leafy stencils above the doors.

Architraves painted in stone soften the contrast between the paler green and dark wood.

Varnish has been used on the doors and dado rail to enhance the natural beauty of the wood.

Wood paneling below the dado line is painted a pretty mid-green that is a few shades darker than the upper walls, echoing the natural fall of light, which is usually brighter higher in the room.

◀ **Old Worlde charm**
Shades of green paired with natural wood combine to make a restful traditional dining room.

◆ **Latex** This modern, synthetic, water-based paint is most often used for walls and ceilings. It is easy to work with because spills can quickly be wiped up with water and a sponge. Latex can be flat or shiny, as you like, coming in matte, silk, and eggshell finishes.

◆ **Gloss/eggshell** Choose a hard-wearing gloss or eggshell for interior and exterior woodwork and metalwork. Traditionally, most glosses were oil based, but they also now come in water-based acrylics, which are more environmentally friendly and easier to clean.

◆ **Primers** These are used to seal bare wood, metal, and plaster, and provide a suitable surface for the paint to adhere to.

◆ **Undercoat** This is sometimes necessary if you want to make a dramatic color change and need to provide opaque coverage over old paints.

◆ **One-coat** This paint needs no undercoat and only one top coat. It is available as latex or gloss.

◆ **Distemper** Soft distemper is a powdery traditional paint often associated with cottage walls. Nonwashable, it is rarely used nowadays, although some traditional paint manufacturers still make it. Casein distemper has a tougher, more wipeable finish and comes in wonderful traditional chalky colors.

◆ **Scumble glaze** A transparent glaze that is applied over or mixed with artists' colors to create paint effects such as marbling, dragging, and stippling.

◆ **Special effect paints** These include pearlized and metallic paints and translucent washes, which are applied to wood to enhance the grain.

◆ **Varnish** An oil- or acrylic-based translucent finish designed to show off the natural beauty of wood.

COLOR COMBINATIONS:
neutral, creams, and whites

Always elegant, neutrals, creams, and whites have timeless appeal, yet they do not have to be seen as safe. Loved by designers for the way they reflect the light, these pale shades highlight beautiful architecture, both clean modern lines and intricate traditional detail. Unwritten oral wisdom dictates that you should never mix white and cream. However, in practice, it can be very successful, enhancing woodwork or a focal point in a subtle way.

Neutral colors that echo nature always look elegant. Think of the colors of exquisite bleached pieces of driftwood, old stone, gray beach pebbles, and tree bark in its various forms. All these hues are enduring, easy to live with, and restful on the eye.

▲ **White and cream**
A white top lends a smart finish to cream cabinets. This elegant combination looks good in both traditional and modern interiors.

▲ White with a touch of color

White always looks fresh and clean and can be highlighted by an extra touch of color, reflecting the overall style of the room. A soft green striped throw adds to the simple country appeal of this bedroom.

◀ Pretty white

White doesn't have to mean pattern free. This mainly white scheme has been broken up by delightful floral and ticking textiles. The floral rugs coordinate charmingly with the chair throws for a cohesive look.

COLOR COMBINATIONS:
pastels

The delicate charm of pastels has made them a favorite for centuries. In the classical Roman period, artists used the soft colors to create frescoes, the elaborate wall-paintings popular at that time. Pastels were also favored by Madame de Pompadour, the mistress of King Louis XV of France. Her imaginative combinations of pinks, blues, turquoise, lilacs, and greens were quickly copied by Paris salons of the 18th century. It's a lovely look that has rarely been out of fashion since. The appeal of these pale shades is their mix and matchability combined with the way they bring color to a room without absorbing too much light.

▲ **Green contrast**
Even modern interiors look good in pastels when used in a contemporary way. Here, the basin alcove is painted green to provide a sharp focus in a mainly cream room.

▲ Lemon delight

Light and sunny, shades of lemon always lift the spirits. Here, pastel lemon enlivens a formal dining room, imparting a spacious feel to the whole area.

◀ Sugar almond shades

The pink and blue combination in this traditional Swedish-style room would look well, even in a modern setting, highlighting a color scheme that transcends time and fashion. Pastels always look wonderful together and will team happily if you choose colors in a similar tone.

PAINT

COLOR COMBINATIONS:
mid- and deep tones

Mid-tones are the shades of summer and fall fruits: berries and plums, apple-greens, and the mellow greeny-yellows of ripening pears. Terracotta, earthy sienna, and the spice colors occupy the mid-tones. It's best to keep either to the fruity shades, or to the spices, and to look to nature for "advice" on how to combine them. Be inspired, for example, by the wonderful pink and green combination of a Victoria plum, or the russet tones of an autumn leaf against forest floor moss.

Deep tones create a rich, moody look, and are especially good for rooms used in the evening, such as dining rooms, cozy parlors, or bedrooms.

▲ **Modern solution**
The new way with deep tones is to paint just one wall and leave the rest white, intensifying the color. Here, chocolate brown adds to a plain white bedroom.

▲ Strawberries and raspberries

Several shades of summer berries combine well with cream floors and woodwork. The strong contrast between mid-tones, such as raspberry-red and white, would have been more demanding to live with.

◄ Blueberry mix

Deep blueberry creates a sophisticated, modern interior. Here, the skirting has been painted to match for a sleek finished effect, while the ceiling is the palest of blues to reflect the light. The azure table both enlivens and sharpens the blueberry.

COLOR COMBINATIONS: brights

Spring, summer, and sunny climes: these are the sources and the inspiration for bright color combinations. Be inspired by corals of Mexico; aquamarines of the Caribbean; blues of the Mediterranean, and exotic pink, turquoise, and reds of India, China, and the Far East. These are the birthplaces of extrovert palettes, where colors can clash and still suit your design scheme.

Be careful of copying them too closely if you live in a chillier clime. They work best in strong equatorial climates, where the light visually bleaches out colors. In soft, northern light, too many bright colors can add up to an uncomfortable combination. Try using bright shades in a simpler way, which will still lend a happy ambience to any interior.

▲ Lime green

The sharp tones of citrus green are surprisingly easy to live with anywhere in the world. Here, lime is accessorized with white for a fresh, sunny look. If this is too sharp for your tastes, try painting one of the walls turquoise, which is more gentle.

▲ Tangerine dream

Soft tangerine tones are more forgiving than their more strident orange cousins. Used in tandem with lemon, tangerine will brighten up any room.

▲◀ Bold statement

Bright colors will always be bold, but that doesn't mean they can't be used in a traditional interior. Here, for example, the background of emerald green walls only goes to accentuate the pretty, curvy lines of the bathroom furniture.

◀ Singular solution

If a whole room of bright colors is a little too intense for your tastes, try painting just one wall. Set against white, the color is intensified, yet it is not too overbearing. The result is fresh and modern.

COLORWASHING

This is a traditional method of painting walls and woodwork to create an antiqued look of surprising depth. It has enjoyed renewed popularity in recent years. Such an effect is achieved by using different shades of colorwash paint, building them up layer upon layer. You can make up colorwash by combining one part colored latex with four parts water. Roughly apply it over white walls, randomly working the brush in all directions. You can also use a pale colored base, then apply a stronger color over it. Three-color colorwashing offers even greater depth, and is particularly effective with a pale final coat that leaves a mist-like layer.

Wood, too, can be colorwashed, using a commercial product or diluted latex in the same way as walls. Paint the wash straight onto bare wood and finish with a nonyellowing water-based varnish.

◀ **Sky blue pink**
This pretty bedroom wall was painted with a first coat in palest pink. Once dry, a light blue shade was colorwashed over the top for an evocative look resembling a summer evening sky. Experiment on wallpaper lining first.

▲ **Spice tones**

Colorwashed strong tones give more depth to the finished look than does a single coat of latex paint. Here, terracotta has been colorwashed onto a sunny yellow base for a warm, textured look.

▲ **Limewash look**

Limewashing whitens wood while allowing the grain to show through. However, since the ingredients of traditional limewash are highly toxic, ask your paint store for a limewash style product or use a white woodwash instead.

MODERN MOVEMENT

It is not difficult to create bold, modern patterns with paint. First apply a base coat color, then draw simple shapes and motifs, and, once happy with the overall design, paint them onto the chosen surface. Many people, however, find geometrics easier to manage as they can be painted neatly with the help of masking tape.

Stripes, checks, squares, triangles, rectangles, or large blocks of color can be painted onto the wall for an interesting up-to-date look. Paint them onto just one wall, or paint them all over, to highlight a focal point, or simply to add interest to a dull room. First paint the wall in the chosen shade, then use masking tape to create the geometric shapes and brush on the paint. When nearly dry, peel off to reveal a perfectly straight line. The trick is to not overload the brush and to brush from the outside toward the center to try to prevent paint from seeping under the tape. Peel off the tape within 24 hours, or it can become difficult to remove.

▶ **Smart stripes**
By painting (rather than papering) these stripes into position, the top stripes can be painted on to create a tented look. Prepare well for stripes, planning them so that they are positioned pleasantly at doors and windows. This may mean widening or reducing all the stripes for a correct fit.

▶▶ **Fluid organics**
This design was painted in layers. Once the base coat was dry, the shapes were drawn on in pencil. First, the loose vertical sections were applied using a rag. The organic shapes were then completed and allowed to dry before the spots and dots were added.

BUILDING FINISHES

Modern interiors are becoming less and less embellished, and this trend affects color. Often it is added at an earlier stage in the building process, rather than painted on at the end. Plaster is being left in its raw pink form, or sometimes polished for a marbled effect. Concrete is pigment-stained for interiors; plaster-like stucco that is traditionally used both inside and out is now being stained before application. The result is a more robust finish that, rather than covering it up, expresses the building process.

▲ Pigment-dyed plaster

Here, the builders mixed a solution of raw sienna into the plaster before application. The result is an evocative terracotta-style finish.

▲ Polished concrete

Ordinary gray concrete has been polished to a sheen, expressing the building and simultaneously giving it a robust, tactile, and smooth finish.

▲ Beaten plaster

This pale plaster has been beaten with chains to give it a distressed look for texture and depth.

▶ Stained stucco plaster

Stucco is a fine traditional plaster designed to be used indoors or out. This layer of stucco has been pigment-dyed in gray to give the wall a rugged concrete feel that is smooth and sophisticated.

PAINT EFFECTS

Faux, the French word for false, is the interior design name for using paint to make quite ordinary materials appear grander. Woodwork as well as plasterwork can be transformed with the clever use of paint to look like sophisticated marble, sandstone, terracotta, verdigris, wood grain, or gilding finishes.

Some effects, like marbling, demand a steady hand; others require special tools or other materials. Plan these carefully as the time and effort either you or a professional take to create them will mean you won't want to be painting over them in a year's time. These methods take time to perfect, so if you'd like to try them, it would be best to first consult a specialist book or enroll in a course. But here's an idea of what's involved.

▲ Terracotta

This effect is created using a colorwash method (see page 64) and by sponging on successive layers of dark yellow, terracotta, and cream latex.

▲ Sandstone

This bedroom has been given a baronial feel with faux sandstone. First, the sandstone shapes were drawn out on a white latexed wall. Next, the wall was given a three-layer colorwash, and finally the faux mortar was hand painted on in pale-gray latex.

▶ Gilding

The faux method of gilding is to apply a layer of gesso and then use gold size to glue metal leaf into position. Gilding is best used for highlighting and detail, rather than on whole walls.

STENCILING AND STAMPING

Quick and easy, stenciling and stamping are an effective way of adding pattern. Traditionally, many wallpapers were decorated with hand carved wood stamp designs, though today this look can be recreated more easily using bought rubber stamps. You could also make a stamp by cutting your own design into linoleum.

Stencils can be used to create rather more elaborate wall motifs because they can be applied in several parts: each one in a different color. Either cut your own using stencil card, or choose from ready-made stencils available in decorating shops. Apply with a stencil brush, which prevents the paint from leaking under the card.

▶ **Stenciling**

A popular method since Victorian times, stenciling is an easy way to transfer both simple and complicated designs onto the walls. Here, two contrasting floral designs are used to create a pretty, coordinated effect.

▶ ▶ **Stamping**

Stamps are usually simple single motifs. The paint is applied to the stamp with a brush or roller and the first stamp made on spare paper to soak up any blobs. The stamps are then printed either randomly, as on this door, or in a more orderly fashion to highlight focal points.

HAND PAINTING

You need artistic flair to attempt painting designs and images directly onto walls by hand. These can range from simple motifs to full-blown trompe l'oeils. There are many professional artists who will create wall paintings for you. Consider carefully the color you'd like for the overall room, and have that freshly painted onto the walls before either attempting or commissioning a hand painting. Most hand paintings are given a place of honor on important walls where they're most likely to be seen. Sometimes, however, entire rooms are decorated with fantasy themes. This is particularly popular with children who love fairytale, underwater, jungle, and space age motifs.

▲ Trompe l'oeil
Lavish swagged and tailed drapes, complete with pineapple pulls, have been painted onto the walls of this period house to complement the intricate cornices. With trompe l'oeil drapes as "soft furnishing," the windows have been left unadorned to let the light flood through.

▲ Permanent art
Painted onto faux terracotta, this image has cave-like appeal, though its simplicity has an exquisite modern look. It works well in a simply furnished contemporary studio and the choice of colors reflects the terracotta background and other earthy colors in the room.

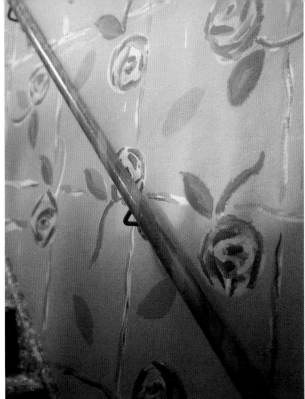

▲ Simple motifs

A stairway takes on colorful contemporary decoration with simple hand painted rose motifs. First, the walls were sectioned into squares, then painted aqua all over. Next, the roses and leaves were painted on, and finally the squares painted around the roses.

▶ Scenic art

This delightful pastoral scene has been painted onto parchment-colored walls. The foreground trees were rendered in stronger tones than the soft village and field background to give a feeling of distance and depth. The whole might otherwise have been too overwhelming.

WALL COVERINGS

elegant

pretty

coordination

texture

sophisticated

smart

elaborate

fine

fun

modern

creative

DESIGNS FOR WALLS

If you want pattern on your walls but your talent with the paintbrush ends at rolling on latex, consider hanging wallpaper instead. There's a wide choice of coverings featuring painted scenes, modern geometrics, and stamp effects, as well as traditional florals and trellises. They can be put up in a fraction of the time that it would take to paint them. One advantage of paper over special paint effects is that you can see the finished look before you buy. There is no need to guess the end result of, say, colorwashed terracotta over sand. Wall coverings can also add texture to the walls—either because they are textured papers or vinyls, or because they are paper-mounted fabrics, such as hessian.

When wallpaper has a large proportion of white or a pale background, even strong patterns like this can make a bedroom look fresh and pretty.

▲ Soft surroundings

Pale pink sprigs of flowers trailing over a white background make a perfect bedroom wallpaper. The pillow covers and lampshades have been chosen to create a pretty, coordinated place to sleep and rest.

▶ Pattern perfection

Traditional wallpapers often feature pastoral floral and foliage designs. Set against pale backgrounds, they offer an easy-to-live-with scheme.

With matching wallpaper and fabric, the windows visually become part of the walls at dusk. The result is a neat, seamless look.

YOUR CHOICE

◆ Lining paper

Use this to prepare the walls for any wall covering. It provides a clean, smooth surface for the top covering to adhere to. Hang it horizontally so the seams don't clash with the vertically hung decorative wallcovering.

◆ Wallpaper

A traditional favorite, wallpaper is still the most popular type of wall covering. The heavier the paper, the better the quality. Designs are either machine printed or hand printed.

◆ Washable and wipeable

PVC-coated paper is wipeable but not scrubbable— a resilient choice for young families where walls are prone to be marked by sticky little fingers. Vinyl is a wipeable, moisture-resistant covering that is usually paper backed. It is a kitchen and bathroom favorite.

◆ Blown vinyls

Some vinyl coverings have deep embossed patterns for extra texture.

◆ Paintable textures

These include embossed papers made from linseed oil and flax fused onto paper used for hard-wearing areas, such as below the dado rail in halls. Once put up, they can be painted and repainted over the years.

◆ Textile wall coverings

These range from paper-backed hessian to traditional flock, which has a pattern of tiny carpet-like tufts.

◆ Photographic reproduction

With electronic imaging, any picture can be reproduced onto wall coverings. Specialists will enlarge the image to whatever size you want and transfer it to wallpaper for easy hanging. Photographic wallpaper looks stunning on just one wall, set off by painting the other walls a plain color.

TRADITIONAL DESIGNS

Florals and trellises, medallions and allover patterns, traditional wallpapers often featured elaborate, multi-colored designs. In the 1970s, there was a huge revival in traditional designs, such as those of the Arts and Crafts period, or pretty sprigged country designs, all harking back to the early 20th century. Many manufacturers produced coordinated wallpapers, pairing a paper with a simple allover design, for example, with one featuring larger, bolder patterns, linked by color or style. The ranges were usually designed with an easy, coordinated look in mind. Nowadays, however, the look is less likely to be fully coordinated and more often a single wallpaper is chosen, in tandem with painted walls.

▲ Arts and Crafts revival

This William Morris-style design is archetypal of the Arts and Crafts movement, with large designs based on nature: foliage, flowers, fruit, and sometimes small animals. These designs give an overall warm, cozy feeling to any room.

▶ Painterly elegance

Some of the most exclusive wallpapers have large repeats that look like a design painted onto the wall. This exquisite design harks back to the European interest in all things Chinese that arose at the beginning of the 18th century.

WALL COVERINGS

TOILE DE JOUY

These distinctive designs were originally printed onto plain Indian cotton furnishing fabric in late 17th-century France and later became a popular design for wallpapers, too. Toile de Jouy is always printed in one color onto a pale background and depicts pastoral scenes of country life—including such images as garlanded ladies in gardens, traveling musicians, riding, hunting, and harvesting scenes.

Toile de Jouy most often comes in pinks and blues, but darker shades are also available, such as maroons, greens, and even black. Because the lines of the designs are fine and there is plenty of white or cream in the background, the overall look is light and pretty—perfect for bedrooms and morning rooms. Matching fabrics are also frequently available.

▲ Toile throughout
Toile de Jouy wallpaper is used to link a corridor and bedroom beyond, visually opening up the whole area. White-painted cupboard doors on the right-hand side have the effect of widening the corridor.

◀ Toile teamed with paint
A red enameled stove could have made an incongruous living room choice, but paired with red toile de Jouy, the overall effect is light and pretty.

▶ In the navy
Toile de Jouy wallpaper in navy blue makes for a pretty bedroom, especially when coordinated with white. Despite the dark ink color, the overall effect is light and delicate because there is plenty of white background.

SMALL MOTIFS

Wallpapers with small motifs have an irresistible charm. They are not overly busy, yet, rather like a colorwash, they have more color interest than flat paint. Small motifs work well in all rooms. They are the perfect scale for small rooms; in larger rooms, they give the overall effect of gentle broken-up color, a spotted or speckled look. Either hang the same wallpaper over all the room, or you could buy two different colorways with the same (or similar) motifs. This works particularly well with a simple spot motif in pastels, for example.

▲ Spotty solution
All small motifs, including spots like these, are the perfect solution for small, awkwardly shaped rooms, such as attic spaces, because there are fewer problems with matching pattern repeats.

▲ Going dotty
Simple allover spots are very appealing. They have a cottage charm combined with a crispness that transcends time. This blue spotted wallpaper has a daintiness that is perfectly suited to a feminine bathroom.

▶ Teacup treat
These teacup motifs introduce a quirky charm to the walls. Each one is an exquisite illustration in miniature, so they don't overpower the room. The border featuring the same teacups on a larger scale echoes the overall design.

ALLOVER TEXTURE

Whether they're printed paint effects or abstract designs, wall coverings that have an allover texture all share the same advantages—they're much quicker to apply than a paint effect and you can try-before-you-buy. Unless you're practiced in these techniques, it can be difficult to imagine the end result of any paint effect. Even if you try out the effect on a spare piece of wood or wallpaper lining, you'll never manage to recreate exactly the same effect as your sample. With wallpaper, you can buy just one roll, pin it up in the room you are decorating, and live with it for a few days before committing yourself.

▲ **Colorwash cheat**

Wallpaper printed to look like colorwash goes up much more quickly than the brush-painted version. This yellow paper adds a sunny feel to a traditional home.

◀ Random life

These random spots, reminiscent of snakeskin, give a sophisticated textured look to the walls. In elegant pink, the wall covering makes for a bathroom that is reminiscent of late 19th-century Europe.

▲ Cool crackle

Faux crackle glaze has the sophisticated look of translucent antique porcelain, but it requires a certain skill to apply correctly. To assure the elegance you're looking for, hang up a cool, neutral wallpaper, such as this one.

WALL COVERINGS

GEOMETRICS

Smart and elegant, geometric designs on the walls have been a popular choice for hundreds of years. Stripes, particularly, became a real favorite in early 19th-century Britain, where they are still known as Regency stripes and where their popularity endures. Generally, the smaller the room, the narrower you'll want the stripes. Hanging geometric wallpapers requires careful preparation, both to ensure that the placement of motifs is attractive in relation to the features in the room and so that they are absolutely vertical.

▲ Wide open

Widely spaced stripes demonstrate the clear need for careful planning in relation to the architectural features in the room. Here, papering has been started at the window, which is framed by smart olive-green stripes, and then carried through along the wall. The effect would not have been nearly so dramatic if the stripe had fallen halfway along the window.

▲ Smart stripes

By painting (rather than papering) these stripes into position, they can be brought over the lower part of the cornice, and the cornice color can be matched exactly to the walls. The pink and buttercream combination gives a light background for the richly colored fabrics of the scheme. They'd look equally well with paler furnishings.

◄ Modern moves

The newest looks for wallpapers have moved a long way from pastoral designs and traditional geometrics. This paper features squares of very similar colors but with completely different finishes—ranging from shiny metallics to matte suede effects.

FABRIC WALL COVERINGS

Fabrics have a long history in wall decoration. In Europe, tapestries were hung on medieval walls, Persian rugs embellished Islamic walls, and silk-tented ceilings have been in and out of fashion since the Arabian Nights. As well as being used as embellishments, fabrics are used as wall coverings themselves. Some are paper backed; others are designed to be mounted onto special battens and fixings to create the appearance of a fabric wall. Most fabric wall coverings have a textural quality—hessian, silks, slubbed cottons—making them wonderfully tactile. Fabric wall coverings imply expense and so offer a sense of luxury.

▲ Two ways with fabric

Striped fabric "awnings" add up to an elegant solution, lending importance to the beds and a sense of dynamism to the walls. It works especially well because the walls have been painted in a matching shade of blue.

▲ Wall to wall

Panels of fabric fixed into special battens on the wall create a wonderfully luxurious look. They're not the best choice when within reach of small dirty fingers, but work well in more formal rooms.

▶ New dimensions
A richly lined bed canopy in an inviting golden bedroom hints at the tented rooms of medieval castles. Here the canopy has been lined in a contrasting fabric and the finished piece has been suspended over a wooden frame.

TILES

shiny

matte

mosaic

waterproof

kitchens

bathrooms

colorful

sleek

creative

smart

country

ARCHITECTURAL FINISH

Waterproof, hard-wearing, decorative: it's little wonder that tiles are the universal favorite wall finish for wet areas, such as kitchens and bathrooms. The Romans and Turks chose ceramic tiles for their baths and, more than 2000 years later, we're still choosing them over any other material. Tiles come in a huge variety of materials and patterns, some of which are arranged in kinds according to use: edging tiles, field tiles (the name used for the main body of tiles), and the occasional patterned tile to go with the less ornate.

Tiles can be plain, patterned, painted, or printed and range in size from tiny mosaics to huge slabs up to 2 feet wide. They can be put up diagonally, in brick formation, or in neat rows. Grout can be white, gray, or colored, producing different effects with the same tiles. The choice can be bewildering, and it is not one you should make in haste. Tiles are expensive and, once up, become part of the architecture. Select them with the long term in mind. Choose classic designs that will endure several changes of paint and wallpaper designs as fashions—and your tastes—change.

YOUR CHOICE

◆ **Glazed ceramic**
These are the most common type of wall tile, made with a gloss or matte finish. They can be printed, which was very popular in Victorian times, or plain; they can be embossed for texture, or even made in relief: a popular traditional style is to have relief rope or fruit patterns, such as grapes or apples.

◆ **Vitrified ceramic**
These have a matte finish and have been baked at a very high temperature for a very hard finish. Most come in plain, restrained colors, though in decades past, they may also have incorporated pattern.

◆ **Glass**
These usually come as mosaic tiles, though more are now being supplied in standard sizes. Bear in mind they will look quite different once in position as the light will not be able to pass through them.

◆ **Limestone**
These can either be matte or glossy and can be supplied up to 2 feet wide for a sleek finished effect. Tumbled limestone are attractive stone-colored tiles that have been processed for an aged look with more rounded corners, and often slightly irregular sides.

◆ **Clay**
European-style quarry tiles are not glazed and usually have a semimatte finish. Mexican equivalents are often waxed for a glossy finish.

◆ **Mirror**
These come in a variety of sizes and, since they reflect light, can make a space seem larger than it is.

◆ **Marble**
These are cut to generous sizes so the marble's veining can be appreciated. They are more easily transportable and easier to attach than slab marble.

Navy blue edging tiles of varying sizes have been incorporated into the mirror surround, picking up on the color theme of the mosaic tiles.

Mosaic tiles in matching blue and white have been used to make up a Greek key style border.

Six-inch white tiles form the background of the design.

◀ **Simple pleasures**

Pale blue mosaic tiles have been used not only for the splashback behind the basin but also for fronting and lining the clever storage area beneath.

▲ **Custom border**

The tiles used in this bathroom come in several different sizes and colors, so they can be combined to create individual designs.

SIZE AND SHAPE

Tiles range from tiny mosaics to huge slabs of up to 2 feet in length. The larger the size of the tile you choose, the sleeker the end result: mosaics bring more pattern interest, even if due only to the grouting. Most tiles are square, but rectangles and hexagons are also readily available. Recently, new shapes have come onto the market including round and random pebble forms. Most tiles are supplied individually, but mosaics of all shapes are usually supplied on a backing of netting, ready spaced for even grouting so they can be put up in sheets, rather than by individual tile.

▲ Rectangular beveled
White brick-shaped tiles like these became popular in the early 20th century. They have been topped with long, narrow tiles, also in white. The beveling accentuates the sheen of glazed ceramics.

▲ Rectangular mosaic
Tiny rectangular mosaics in different shades of cream applied vertically make an interesting modern chimney feature. The pale tiles are offset with darker taupe grout to add to the textural effect.

▲ Mosaics

Any small tile, which is usually square and made from glass, can also be made from glazed or vitrified ceramic. Mosaics can come on a sheet for quick application or can be used individually to create a Roman-style intricate design. They also come in a variety of different shapes, such as round or pebble form.

▶ Clever edging

Classic beveled rectangular tiles can be neatly finished off with special edging tiles. These are simple quarter round profiles, but there are many styles available, including elaborate textural fruit designs.

PATTERNS

Patterned tiles featuring elaborate designs screen printed onto the surface of the tile before baking became extremely popular in Victorian times. The rich colors that are associated with Art Nouveau tiles were incorporated in the glazes, while the simple, brightly colored designs on Mexican clay tiles were often hand painted.

Some tiles have an allover pattern, while others are part of a whole, each contributing to a final design or even a picture. You may have patterning only in the corners, which combine to make a different design when the tiles are fitted together. Another alternative is to set a wall of plain tiles and randomly add a few decorated tiles for visual interest.

▲ **Picture story**
Screen printing allows for the manufacture of elaborately decorated tiles, such as these flower baskets. Such ornate designs make an attractive focal point.

▲ Printed team

Screen printing is a relatively inexpensive way of producing patterned tiles, especially if it uses only one ink color. These tiles printed in blue come in a coordinated "family." Here, the "field" tiles covering most of the walls have been printed, but only in the corners. When combined diagonally, they form a simple allover motif design. The coordinating border tiles are printed likewise to bring emphasis to the overall design.

◀ Random motifs

Some of these tiles have corner decoration only, while others feature a large motif to allow for design flexibility when they're fixed in place. Here, every other tile has a motif. The images have a hand-painted look and, in times past, they would, in fact, have been hand painted before they were fired. Today, however, they are more likely to have been printed.

LAYOUT

Even if you buy simple square tiles, you can create different effects, depending on whether you put them up in straight lines or set them diagonally; and whether you match the grout to the tiles or go for a contrast. The same tiles, when differently colored, can also be used to make various (usually geometric) patterned effects. Using tiny mosaics, create more elaborate images over a large area, such as checkerboard, striped, or banded effects.

▶ **Checkerboard style**

Here, red and white tiles in the same size and from the same range have been used to create country-style checks in a kitchen.

▼ **Tapestry tricks**

A range of geometric tiles in muted earthy shades have been randomly arranged to create an interesting tapestry-like design.

FLOORS

FLOORS
INTRODUCTION

Careful thought needs to go into the choice of floors. Hard floors, such as stone, ceramic, and wood, especially, are long-term choices that are essentially part of the architecture. They are expensive, require expert laying, and for both those reasons their choice should not be influenced by current fashions or trends. These are floors that should be expected to last decades, if not for the whole lifespan of the house, and should be chosen as team players that work well with changing fashions in wall coverings and furnishings.

Since hard floors are usually made from natural materials, longevity is usually not a problem. The best solution is to choose flooring that is available locally as it will always endure fashion changes. The fashion element in floors can be introduced with rugs laid on top of hard floors. If you desire frequent change, choose soft flooring, such as carpeting, or semihard flooring, such as vinyl or rubber, which are more easily replaced than their hard counterparts.

▲ Stone tradition
Flagstones are an exquisite flooring that dates back many centuries in Europe, and still looks good today. The most successful stone floors are made with locally quarried stone, as they link the house with its natural outdoor surroundings. When polished, the sheen is beautiful.

◄ All white
White painted floorboards have a long tradition, in both Sweden and New England, and their light-reflective quality is enduringly attractive. You need to use special floor paint, then give it extra protection with heavy duty varnish to create the look that is currently popular. If you like, it can easily be changed back, by sanding and sealing, to give the more classic natural wood look.

▲ Wild at heart

Animal skin rugs transcend fashion because of their natural patterning and colors, although most people nowadays would prefer to buy fake versions instead.

◀ Simple style

Woven cotton rugs, such as Indian dhurries, are the ideal way to bring a fashion element to hard floors. Flat woven in cotton, they are available in an infinite number of designs, ranging from simple stripes, like this, to intricate geometric and floral patterns. They are relatively inexpensive and easily changed.

STONE & CERAMIC FLOORS

enduring

timeless

hard-wearing

marble

natural

slate

flagstones

quarry

permanent

tiles

TIMELESS STYLE

Stone of any kind has been a favorite flooring since ancient times. Because it wears well underfoot, it should, ideally, last for the lifetime of the house. In some parts of Europe, this can run into centuries, with the only sign of deterioration being a gentle wearing of the stone in the areas of hardest traffic, such as door thresholds. Archeologists still unearth exquisite Roman marble floors, relics that date back some 2000 years. The permanence of stone elevates it in status above decorative floor coverings and makes it an integral part of the architecture. It should never be chosen as a fashion statement. The most suitable sone is quarried locally, putting the house in context with its surroundings, and providing a true sense of belonging.

▲ Limestone elegance
The pale tones of natural limestone give a contemporary feel to a lasting floor. These have been set diagonally on the kitchen floor both to enhance the modern look and to lend a sense of space.

▶ Flagstone finesse
Flagstones, which are usually cut from local granite, generally give a down-to-earth feel to farmhouse kitchens. Here, the interesting way in which these flagstones are laid and the addition of a tessellated (mosaic) border provide a greater sense of sophistication.

YOUR CHOICE

◆ Marble
An exquisite stone with delicate veining that can be polished to a high gloss. Extremely hard, this is seen as the ultimate stone finish for interiors.

◆ Granite
In hardness, this is second only to marble. It is available in a wide range of colors, from cream to almost black, and can be speckled, lined, or spotted.

◆ Limestone
This has a fine texture and comes in a range of sandy, gray, and almost brown tones. It is not as hard or resilient as marble and granite. Limestones are often named after the region in which they were quarried, such as Portland stone and York stone.

◆ Slate
A hard stone that most commonly comes in charcoal grays, but it can also be dark green, blue, or brown. It is characterized by a horizontal splitting that gives it a naturally nonslip, riven surface.

◆ Travertine
A hard, marble-like stone that comes in exquisite soft tones, from cream to buff browns, which often has a subtle banding.

◆ Clay tiles and bricks
These are made from quarried clay that has been shaped, then baked at a very high temperature to create a strong, resilient building material.

◆ Ceramic tiles
Any fired clay tile that may or may not be glazed. Ceramic tiles are most usually square, but you can also find triangles, hexagons, rectangles, or lozenge shapes that can be laid as mosaics.

Although all the flagstones are the same size, they can be laid in different directions to delineate different areas within a large room and to emphasize circulation spaces. Here, the central part of the room is laid diagonally, while the circulation areas are set square on.

A border of tessellated floor tiles is used to create a boundary between the square and diagonally laid flagstones.

MARBLE & GRANITE

The hardest of all stones, marble and granite are both igneous rocks, which means they were formed when limestone was put under extreme heat and pressure, producing a crystalline quality. Marble is generally characterized by fine veining, while granite contains sparkling quartz and feldspars, often forming elaborate allover patterns. Igneous stones are heatproof, water resistant, easy to clean, and durable.

▼ Natural pattern
Many granites have this attractive allover speckling. Some stones also contain quartz and feldspars, which add sparkle to the finish.

▲ Cloud effect
The ethereal cloudy effect of this soft gray marble is preserved by the way it is cut and laid. Large long sections are cut at the quarry from a single slab of marble. Then they are re-laid in consecutive order just as it was before being cut to retain the natural pattern of the stone.

▶ Black beauty
Characteristic veining is one of the hallmarks of marble. This black marble with white veining has the reverse effect of the more classic pale gray coloring.

STONE & SLATE

Limestone is softer than marble and granite, but its even, creamy tones make it a popular flooring. Because it is more porous than igneous materials, it is less durable and needs a protective finish to prevent staining. Formed over millions of years from layer upon layer of shells and bones of marine creatures settling on the sea floor, it comes in sandy tones, grays, and browns, and is often faintly patterned by the fossilized remains of sea creatures. Slate is formed under heat and pressure from sedimentary rocks, which creates its distinctive dark riven features. Inexpensive, impervious, and durable, it is a handsome, hard-wearing flooring.

▲▶ Tumbling stone

Tumbled limestone has a more textured finish with country appeal. However, even this less formal style limestone needs a finish to resist stains.

▶ Extra wear

This dressed limestone has been given a protective coating, which resists stains and marking.

▶▶ Smart slate

With its deep shades of gray and its durability, slate is a wonderful flooring—beautiful to look at and easy to care for. It is also one of the most inexpensive stones.

CLAY TILES & BRICKS

Clay tiles are made from baked clay, rather than quarried stone. Fired at high temperature, they are resilient and hard-wearing, even when unglazed. Quarry tiles and clay tiles are usually made from unglazed local clay and are fired at high temperature using a method similar to brick making for strength. In Europe, they are usually square, though tessellated (mosaicked) floors are a popular tradition. These are made up of various small geometric shapes, such as triangles, squares, and lozenges, and in colors ranging from creams to tans and near blacks. Mexican equivalents come in various shapes including lozenges and hexagons. Quarry tiles have a rich natural patina, which improves with age.

Bricks can also be used for flooring, bringing a reddish shade to the floor similar to that of clay tiles, but offering a variety of different laying patterns.

▲ Brick beauty

The simplicity of bricks can bring surprising style to floors.
Their neat size and shape offer plenty of scope for
different laying patterns, adding a sense of originality. The
complicated shape of this roof space is pulled together
with its diagonal brick-laid floor.

◀ Pattern to delight

These rectangular clay tiles have been made more
interesting with an impressed motif. It creates a floor that
is restrained yet delightfully patterned.

◀◀ Aged to perfection

Large quarry tiles laid on a farmhouse floor and regularly
polished will mellow beautifully. They will take on a rich
patina as they age.

GLAZED CERAMICS

For color and variety, nothing can surpass glazed ceramic tiles, which come in the same styles as wall tiles (see page 94), except that they are more heavy duty. Ranging in size from tiny mosaics to tiles up to 2 feet across, glazed ceramics come in a rainbow of shades. And, just as with walls, the design scope is wide indeed. They can be set square-on, diagonally, or brick bond style (if they're rectangular). Or, you can use a variety of geometric shapes to create mosaics.

The choice of grout can affect the design, too. A toning grout harmonizes with the tiles for a restrained look, whereas contrasting grout adds impact by delineating the tiles.

▲ Mosaic underfoot
For an allover look, choose mosaics, which work especially well in small spaces, as seen here. The grouted lines create an integrated pattern.

▶ Classic contrast
Although this checker design presents the starkest of contrasts, it is not overpowering as its traditional roots give it a comforting familiarity.

▲ Cool simplicity
White glazed tiles make a sophisticated floor for an all-white kitchen. Their robust and stain-resilient surface means they're easy to keep clean and will continue to look smart for many years.

▼ Checker style
Checkers are a classic design, but they need not be restricted to black and white. For more harmonized schemes, choose two colors that are close to each other in shade, such as this cream and taupe.

WOOD FLOORS

natural

versatile

timeless

durable

patina

rustic

woodgrain

strip

boards

parquet

block

NATURAL TONES

Wood comes in a wide variety of magnificent colors and grains. They have universal appeal as they are a continual reminder of nature around us and so team well with almost any interior scheme. Whatever style and color wooden floor you choose, it will visually improve with age, building up a patina from the polishing effect of daily wear and tear. However, wood does not have the resilience of stone and ceramic and must be treated with respect. No wood takes kindly to furniture dragging, or to stiletto heels. Slim-legged furniture can also create problems by causing indentations. Having said that, surface damage can be restored with a light sanding and refinishing.

Shades of wood run from the palest beech to the darkest mahogany and, just as with all interior design, there are trends—one year it is lighter woods, another year darker. But try not to be too swayed: your choice of wood flooring should be seen as a lasting investment and not a fashion choice. If in doubt, be guided by the original wood in your home.

▲ Pale and interesting
The rich hue that traditional floorboards take on over the years works well with any wall color and other woodwork. Here, pale shades are a refreshing contrast to the lovely knotted boards on the floor.

▲ Original style
Restoring the floorboards is always the best solution when refurbishing old houses. Specialists can lift them, cut out damaged parts, and re-lay, closing any gaps that have appeared with age. The floor is then sanded and sealed.

The rich brown tones and elegant graining of walnut add stature to any room, and should always be restored rather than replaced.

The natural walnut color has been enhanced with a toning stain and waxing.

YOUR CHOICE

◆ Bamboo
Although a grass rather than a wood, bamboo can be found at most lumber yards. Resilient and durable, it resists moisture and so is used in bathrooms.

◆ Beech
A reddish fine-grained wood with more grain interest than birch.

◆ Birch
A very pale hardwood with marked knots of varying sizes, but otherwise subtle graining.

◆ Cherry
An attractive, rich, glowing wood with pink-red tones. Also available in golden yellows.

◆ Elm
This is a beautifully grained mid-brown wood with reddish tones.

◆ Maple
An extremely hard, fine-grained wood that is creamy white, but also available in darker browns.

◆ Oak
Neither too dark nor too light, oak's mid-tone shade means it works well in most situations.

◆ Pine
Pine is very popular, admired for its lovely pale yellow color characterized by knots and an open grain.

◆ Teak
A magnificent deep red-brown wood from the tropics, teak is extremely hard and fine grained.

◆ Walnut
A mellow, dark grayish-brown wood with an elegant and especially interesting grain.

BOARD STYLE

Choosing a wood floor involves far more than simply picking your favorite color and grain. Wood flooring is presented in many different forms to suit all budgets, situations, and design tastes. Start by visiting the showroom, taking home samples, and deciding which look you like best. At the same time, you should consult with a floor fitter who will advise you on what flooring would suit both your needs and the architecture of the house. Not all wooden floorings are suitable for all situations, and this can depend on the existing sub-floor.

Once you've chosen the wood you want, you'll have to select a board width and grade of lumber. Most lumber yards offer four grades ranging from rustic, which is knotty and has plenty of color variation, through select and factory grades to first grade, which should have no knots and no color variation. Again, the fitter who will lay down the floor can advise on both width and grade.

▲ **Strip show**
Strip oak flooring with its mellow color and distinctive fine grain with few knots, makes any room look good.

▶ **Board walk**
Wide boards are the most traditional wooden flooring. Boards wider than six inches need to be screwed down, though the screw holes add to the character.

◆ Strip flooring

These are narrow strips of solid wood of varying lengths and with tongue and groove on all four sides. They can be laid straight onto joists or battens or laid over an existing wooden floor.

◆ Boards

These are wider and longer than strip flooring and have a more traditional look. If the boards are tongue and grooved, they will have to be secret nailed. If tongue-and-groove boards are wider than six inches, in addition to being secret nailed, they will also have to be screwed and plugged through the top.

◆ Prefinished

This flooring come as strips or boards, which are supplied ready finished from the factory, including a protective coating. You will find a range of finishes, from high gloss to semigloss polyurethane for a contemporary look, or burnt, wire brushed, and oiled, for an antique look.

◆ Engineered or laminated

Two or three strips of prefinished hardwood boards are laminated side by side onto a plywood base. The result is a stable flooring that is less likely to warp than one made from traditional boards. It is also flexible when it comes to fitting as the lengths can be sawn to fit any space. The timber can also be fitted as a "floating floor," which means the boards are glued together and fixed to the walls, not to the existing floor.

◆ Plastic laminate

Although this is not real wood, it can look very good indeed, and can certainly be more resilient than the real thing. To make it look realistic, a high-quality photographic print of the desired wood is bonded onto a high density base, which is often moisture resistant. A special coating is then applied over the photographic layer to protect it.

WOOD BLOCK & SPECIAL EFFECTS

Solid wood flooring is most frequently laid as simple strips or boards. However there are other options—intricate herringbone or brick bond styles called parquet. These elaborate designs are beautiful, but they can be time-consuming to install. Rather than lay the wood on site, many manufacturers will supply the more intricate parquet flooring designs on woodblocks. These are square base "tiles" made up of small strips of wood arranged mosaic fashion. They make laying quick and easy. Each choice of design can be laid in a variety of ways for different visual effects. Another option is to lay simple strips or boards and then set off the floor with special borders. Or, you can have a medallion made of an inlaid contrasting wood as a focal point. Both borders and medallions can be purchased ready-made or custom-made to your specifications on site.

▲ Classic herringbone
This exquisite oak floor has been laid piece by piece in a herringbone style for a handsome, timeless look.

▶ Blocks and lozenges
These simple blocks of parallel pine strips have been inset with black painted lozenges that fit together to border each one. The overall effect is of an unusually shaped checkered floor.

▶ Inlaid style

A giant star is the central motif of this intricate marquetry panel. It was specially designed to suit the particular proportions of the room.

▶▼ Giant herringbone

These prefinished machined boards are made up of diagonally laid pine strips. When they are positioned in alternating directions, they create a large and unusual herringbone design.

FINISHES

All wood floors need a finish to protect them from scratches, gouges, and stains. Traditionally, floors were finished with a penetrating stain followed by wax, which provided a low-sheen protective seal. However, waxed floors need regular polishing with a solvent-based wax or buffing paste. Because of their high maintenance, waxed floors have fallen from favor in recent years. The more popular modern solution is to enhance the natural wood grain and color with a translucent gloss or semigloss finish that protects the wood while giving it a light-reflecting shine. Choose from solvent-based polyurethane or the more environmentally friendly water-based varnishes. For high wear areas, such as halls and stairs, consider a two-part epoxy resin finish, which needs professional application.

Striking effects can also be created with finish. You can add color with floor paint, though bear in mind most is made in Scandinavian countries where it is normal to remove shoes indoors. "Wear time" for painted floors is likely to be shorter in American homes, where people prefer to remain shod. Floors painted with country-style patterns and then varnished were popular in Colonial times, and they still provide an attractive solution that is less high maintenance than carpeting. To deepen the color of the wood, you can also use a dark colored stain followed by a translucent finish.

▲ White solution

Tatty original floorboards can be given an attractive new look quickly and inexpensively with a splash of white floor paint and a coat of translucent semi-gloss varnish. Alternatively, use three coats of a paint suitable for wood and finish with two coats of varnish.

▶▲ Fine stripes

This oak floor was sanded before being painted with decorative black stripes. Finally, it was given a coat of high gloss varnish for protection.

▶ Stained pattern

A country-style rug design, painted onto the floor with colored stain, is set off by black paint around the perimeter.

◀ Checkered floor

Paint does not have to stay plain: you can use it to make any pattern you like. These checkers were created using masking tape to ensure straight lines. Once painted, finish with varnish as above.

RESILIENT FLOORING

vinyl

rubber

cork

linoleum

cushioned

easy

quick

durable

washable

diy

clean

EASY LIVING

Water resistant, easy underfoot, often inexpensive, and easy to lay, resilient flooring became the fashion choice for kitchens and bathrooms in the 20th century, but rarely ventured into living, dining, or bedrooms. Today, they are still popular. Because resilient floorings are a much less permanent solution than stone, ceramic, or wood, they can almost be changed along with the interior color scheme. Vinyl, especially, is a do-it-yourself favorite, sold by the yard to be laid with the help of little more than a very sharp knife for cutting. Other resilient products may not be so do-it-yourself friendly, but they come in a wide range of colors and designs, which keeps them in demand.

▲ Textural teaming

A less than smooth surface has been chosen for the resilient flooring in this room to accompany the textured plaster walls.

When a vinyl or linoleum flooring is being laid, careful attention to detail for positioning needs to be kept in mind. Here, whole squares butt up to the island worksurface.

Resilient flooring is made in a wide range of designs. Some of them are very good replicas of natural products, in this instance stone.

◆ Vinyl sheet

Available in plain colors or printed, often with tile-like designs. Very easy for nonprofessionals to cut and lay.

◆ Vinyl tiles

These fall into two categories: the less expensive is similar to sheet vinyl, but cut into tiles, which some people find easier to lay. At the top end of the market are hard vinyl tiles, which come in a wide array of colors and designs. However, they can be almost as pricey as hard floors.

◆ Linoleum

In the mid-19th century this natural floor covering, made from linseed oil, resins, cork, ground limestone, and wood dust became popular. However, it fell from favor when vinyl appeared in the 1960s. Linoleum's current revival is due to new designs and its environmentally friendly components.

◆ Cork

These are really vinyl tiles, but the decorative layer is made from cork rather than printed. The cork-colored surface has a mottled, textural appeal.

◆ Rubber

Rubber flooring dates back hundreds of years. Because it comes in a broad range of wonderful colors it has recently become popular again.

◀ Carefree country

The rust-brown mottled floor in this kitchen lightens the darker background; a lovely contrast to the olive-green kitchen cupboards.

RESILIENT FLOORING

VINYL

When vinyl was first introduced, it was greeted as a liberating new flooring for kitchens and bathrooms. Previously, dark high-maintenance linoleum had ruled supreme, requiring weekly waxing to keep up its appearance. But now vinyl was endlessly colorful and patterned, easy to buy off the roll, and easy to lay. Once down, vinyl was easy to keep clean with just a mop and floor detergent.

The top layer provides cushioning, making it soft underfoot. But not all vinyl is inexpensive. Designer hard vinyl tiles often have a price point close to that of solid hardwood, and each one is custom made to fit the floor. However, they are designed to be long lasting while keeping up with current fashions.

▲ **Mosaic vinyl**
Any pattern can be printed onto vinyl. In this clean, modern interior what looks like tiny white mosaic is, in fact, a patterned vinyl.

◀ Dark contrast

Not all vinyl flooring is smooth. Textured resilient flooring is also available, which is especially valuable in a kitchen or bathroom where water can make it slippery underfoot.

▲ Cool contrast

For a crisp flooring that is easy to look after, vinyl is the perfect solution. The fact that it is easy to clean means that, for once, a white floor needn't be a burden.

LINOLEUM

Invented by Englishman Frederick Walton in 1863, linoleum, made from linseed oil, enjoyed popularity for about a hundred years before falling out of favor. It is now making a serious comeback in Europe, where it is manufactured in a wide range of modern colors with an easier-to-care-for finish. Made totally from natural products, it is environmentally friendly, and improves with age due to the oxidation of the linseed oil. It is also easy to cut, and is often used to create marquetry-style patterned effects. In comparison to vinyl, linoleum is still relatively expensive. However, once laid, it can last for up to 50 years.

▶ Flexible design

Linoleum is easy to cut, which makes it an ideal material for designing special effects. Here, large rectangular tiles are laid brick bond style, the close-but-not-matching shades making up a pleasing design.

▼ Cork

Real cork incorporated into linoleum tiles adds up to a beautiful, natural looking floor. Better yet, it's inexpensive and very easy to maintain.

RESILIENT FLOORING

RUBBER

Rubber flooring currently has industrial-inspired cutting edge designer appeal, further enhanced by the fact that it can be custom-made in any color. It is available with a smooth surface, or with a choice of studs, ridges, or squares and in various sizes, giving a wide choice of different looks. Longer lasting than vinyl, rubber flooring is also more expensive and requires professional laying, which adds to the cost. Available in a choice of tile sizes, it comes with a protective factory finish, which needs to be stripped off once the floor is laid so that a final protective wear coat can be applied. Once down, rubber is smart, durable, and should keep its good looks for many years.

▲ White luxury
Rubber has to be the best solution for a smooth white floor. Easy to keep clean, small stains and marks can simply be mopped up.

▶ Checkerboard style
Contrasting rubber tiles can be laid checkerboard style for a classic kitchen floor.

SOFT FLOORING

comfort

luxury

warm

cozy

deep pile

natural

pattern

carpet

rugs

twist

soft

SOFT FLOORING

CARPETS

The softest of all floorings underfoot, carpet is an enduringly popular bedroom choice. For the rest of the house, wall-to-wall carpeting still holds great appeal: it is relatively cheap and easy to install (though this should be done by professionals). One color can be fitted throughout the house for a smart seamless look and to create a feeling of space. Furthermore, it provides excellent heat and sound insulation. However, since its 1970s' heyday, wall-to-wall carpet has become less popular. It is difficult to keep clean as it ages, despite stain resistant fibers and steam cleaning systems, and the pile flattens over time. Kept to less high traffic areas, such as bedrooms and upstairs landings, however, a good quality carpet should give years of wear.

◀ **New horizons**
Although plain carpets are the most popular modern choice, small allover designs give a fresh spin on pattern and look charming in a bedroom.

Wall-to-wall carpeting provides a sumptuous, soft underfoot choice for bedrooms.

Elaborate patterns offer a pretty flooring for a feminine bedroom.

FIBERS

◆ Wool

This is the carpet fiber of choice: it has good body, is least likely to flatten, and is naturally stain resistant. It is often mixed: 80% wool with 20% synthetic fiber—both for economy and to provide extra resilience.

◆ Acrylic

This is the closest to wool in appearance, though it is not as resilient, and is more inclined to show stains.

◆ Nylon or polyamide

More resilient than wool, but more inclined to flatten and attract stains. Sometimes the fibers are crimped to give the impression of a denser pile.

◆ Polyester

The softest of fibers, often used for deeper pile carpets. Look for the densest possible pile, as polyester flattens easily.

◆ Polypropylene

Hard-wearing and inexpensive, polypropylene is most likely to be used as a mix with other fibers.

◆ Viscose

A cheap fiber that is often mixed with other fibers for bulk.

◀ **Traditional style**

Highly patterned Axminster carpets were first inspired by Persian rugs. Looms were developed to allow for the quick weaving of many colors and intricate patterns.

CARPET CHOICES

After the pattern has been chosen, one must consider where the carpet will be positioned and how much wear it can take. High traffic areas, such as halls and stairs, require the most hard-wearing carpets, followed by living and dining rooms. Upstairs carpeting does not need to be so resilient, so bedrooms and landings can be fitted with a medium weight material. Bathroom carpets should be backed with a high-quality, water-resistant foam. Judge the quality of a carpet by the density and fiber content of the tufts. If you would like the same look throughout the house, choose carpets made up of the same pile but that vary somewhat in appearance.

▲ Smart stripes
Widely spaced stripes are an elegant alternative to plain carpet. This works well because the room has good proportions. Stripes running the length of a narrow room could make the space look even narrower.

▶ Stair runners
Traditionally, stairs were not fitted wall to wall, but with runners like this, held in place with rods. It is still a popular solution, but can be expensive.

YOUR CHOICE

PILE
◆ Tufted and twisted
Tufts are the most common pile, made by punching the yarn into a backing. For twists, the yarn is twisted to make it more hard-wearing and to add extra bulk.

◆ Velvet
A short pile that is dense and hard-wearing.

◆ Shag pile
The pile is cut long for a luxurious feel. Choose only the densest piles as this makes it less likely to flatten.

◆ Loop
The surface is made up of uncut loops, which can be long or short. Sometimes a mix of cut and uncut loops is used to create an interesting textured design. Uncut loops are less likely to flatten than a cut pile.

◆ Berber
Undyed wool carpets with a looped pile. Styles range from paper-backed hessian to flock with tiny tufts.

◆ Cord
Pile made with very short loops to look like corduroy. These carpets are very hard-wearing.

WEAVES
◆ Axminster
The tufts are woven into the backing from above. This way the yarn does not have to run along the underside and many colors can be used.

◆ Brussels weave
An uncut looped pile, usually associated with high wool content carpets.

◆ Wilton
Woven from one strand of yarn to produce excellent plain colored carpets with a velvet pile.

NATURALS

Carpets made from natural fibers bring interest to floors with their wonderful weaves and textures. The main players are jute, sisal, coir, and seagrass. Once relegated to backing, naturals have now become a floor covering in their own right, occupying the limelight in the smartest homes.

The other appeal of natural fibers lies in their easy-to-live-with neutral coloring and their smart hard-wearing looks. Some have a little color woven in among the neutrals for added interest. All come in a variety of textures from herringbones and basket weaves to a whole range of knobbles and bobbles.

▲ **Softest story**
The soft, almost wool-like, quality of jute makes it the best natural flooring for bedrooms. With its creamy tones and textured weaves, jute is visually pleasing and serene.

Grown in paddyfield-like conditions, seagrass is relatively water and stain resistant. This makes it perfect for sun rooms and such, where there is constant traffic in and out of the garden.

◆ Coir

This is one of the sturdiest of the natural fibers and thus the fiber traditionally used for entrance matting. Woven into herringbone and basket weave designs, it provides a handsome and hard-wearing solution for heavy traffic areas.

◆ Jute

Traditionally used for making rope, jute is the softest of the naturals with a lovely slight sheen. Sometimes it is combined with wool for use in bedrooms.

◆ Seagrass

This pleasing weave comes in a soft green and has more sheen than other naturals. Because its surface is slightly slippery, seagrass makes an unsuitable material for stairs.

◆ Sisal

Slightly softer and finer than coir but still sturdy, sisal is a popular choice. It is both smart and hard-wearing.

Coiled into square-shaped sections, seagrass takes on a checkered look.

▲ Style on a shoestring

Although seagrass is an inexpensive flooring, it does not look out of place in a smart interior, such as this. However, it may need to be replaced after a couple of years.

SOFT FLOORING

NATURAL CHOICES

Natural flooring has a crisp modern look that's equally at home in traditional settings. Its neutral tones work well with any color scheme, and it is hard-wearing enough to look good for many years. The downside is the fitting. The natural material is harder to handle than traditional wool and synthetic fiber carpets, and it needs to be secured to an underlay with adhesive. This means that fitters will generally find an excuse not to lay natural floorings. They are worth persuading though. Once down, natural floor coverings look stunning, resist stains, do not harbor house dust mites and moths, and keep their fresh look long after other fibers appear tired.

▲ Pale and interesting
Sisal comes in the widest range of colors, including this pale cream, which looks stunning contrasted against black willow furniture. These palest shades should be treated for stain resistance.

◀ Good economy
Seagrass is generally seen as the most economical natural flooring. But it can look surprisingly luxurious and fits well in the most elegant of living rooms.

▼ Inside out
Twill woven coir is a sturdy yet chic choice for a country living room that opens out onto the garden. Given a regular vacuuming, it should keep its good looks for years.

RUGS: MODERN

If you yearn for a little softness underfoot, yet wall-to-wall carpeting is not your style—and you want to show off a beautiful solid floor—rugs offer the ideal solution. You can make a statement by selecting a rug with a strong design or striking color, or choose a more restrained plain shade to "ground" furniture groupings.

Modern rugs offer a wide choice of looks, ranging from a rainbow of plains to bold bands of color to out-sized geometric shapes. Many are made from tufted wool or wool mixes. But you can also find handsome flat weave, cotton kelim-style rugs featuring simple modern motifs that work well with the latest in interior design, or sumptuous woven chenilles in cream and muted shades.

▲ Smart stripes
Black and beige stripes lend a distinctly modern feel to a traditionally made flat weave cotton rug, which looks good with the wooden floor.

▲ Soft luxury

To add sumptuous texture to a modern room consider using a deep shag pile rug. The contrast with a wooden floor looks and feels wonderful.

◀ Fashion story

Indian dhurries are still flat woven from cotton in the traditional way. However, Western buyers work closely with the local people to come up with designs and colors that suit the modern home. Dhurries provide an effective way to bring an inexpensive fashion element to the floor.

RUGS: TRADITIONAL

Rugs have decorated floors since biblical times, some of the best coming from Persia, now modern-day Iran. Over the centuries, in many countries, different weaves were developed, some flat, some with pile, often depending on the climate in which they originated. In some regions, animal skins and rags were used to make rugs. In others, intricate patterns with religious or cultural significance were designed. Many traditional rugs still survive today, however buying and selling them requires highly specialized knowledge. But there are other options. Many of the traditional weaves and styles have been given modern interpretations, produced with contemporary motifs or in the latest colors. Inexpensive reproductions of traditional oriental designs can also be found.

▲ **Pretty solutions**

A traditional European pure wool rug, woven in soft, muted tones, brings an elegant, yet pretty touch to a wood floor.

◀ **Braided**

Traditionally made from old rags, braided rugs often come in soft, muted shades, imparting a simple country feel to any room.

◆ **Oriental**

Made for century upon century, each rug-making region of the world has its own unique motifs made of knotted wool. Colors range from rich reds and blues to delicate creams and pastels.

◆ **Dhurry**

Inexpensive and colorful hand-woven cotton rugs from India, which are made in traditional and contemporary designs.

◆ **Kelim**

Tapestry rugs from Turkey and the Caucasus, kelims are woven from thick wool into colorful, often geometric, designs.

◆ **Chinese**

Deep pile, colorful rugs featuring traditional Chinese motifs and patterns.

◆ **Braided**

Strips of rags are braided, then coiled from the center outward into oval or round rugs.

◆ **Rag**

These are made from strips of rags that are hooked as tufts into a base made from canvas.

◆ **Sheepskin**

Whole, cured sheepskins, which make beautifully soft bedside rugs.

◆ **Animal skin rugs**

Zebra, cow, and antelope are just a few animal skins that were popular traditionally, but the use of which is now criticized by animal rights groups. Fake versions are available.

◆ **Serape**

Brightly colored woven Mexican rugs.

WORKSHOP
REFERENCE

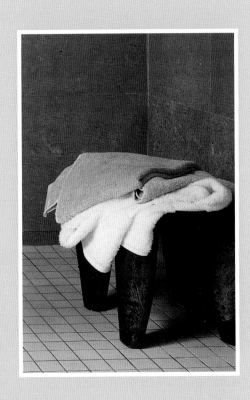

MAKING IT HAPPEN

While redecorating need not be complicated, depending on your choice of finish, laying floors is rarely a job for an amateur. If your walls are in very good condition, then a new coat of paint can simply be a weekend's job, depending on the size of the room. However, the majority of walls are far from being in pristine shape, in which case decorating a room is likely to end up taking considerably longer.

As any decorator will tell you, the quality and longevity of the finish is dependent on preparation. For example:

◆ Loose or flaky paint on woodwork needs to be rubbed down, scraped off, or, in extreme cases, burned off before the wood can be primed and undercoated in preparation for painting.

◆ Any existing latex paintwork needs to be sound, clean and dry.

◆ For a really professional finish, it's best to hang wallpaper lining before painting or wallpapering. (If you plan to use wallpaper, hang the lining paper horizontally to avoid the possibility of seams clashing.)

Of course, most of this is well within the scope of an experienced do-it-yourselfer. Given a good firm floor, laying sheet vinyl and laminate flooring are a fairly simple do-it-yourself job. However, any other floor really does require an expert fitter to ensure you have the best finish. Carpet needs underlay and special tools plus the experience of handling large, heavy rolls. Stone, tiles, and timber need to be laid on appropriate sub-floors, which should be properly prepared and require experienced setting out.

In the right order

Hard floors, especially, are essentially part of the architecture and should be laid even before the walls are plastered. If you're installing a timber floor, the layer should apply only one or two coats of varnish as protection before the plastering. Once the whole room has been decorated, the floor will then be given a light sanding (taking off any errant plaster or paint spills), and the final coat of varnish applied. Stone and ceramic floors should be protected with plastic sheeting until all the decorations are finished. Resilient flooring—such as vinyl and rubber—and soft floors—such as carpets—should be laid after the wall decorations are complete.

If you are decorating the walls and not changing the floors, always ensure the floors are protected with plastic sheeting fixed down with masking tape. Not all contractors show such diligence and may need a little management—so keep an eye open when work commences! Wall tiles should be fixed up once fittings such as baths and kitchen units are in place and the walls decorated. They are then grouted before the edges are tidied with mastic.

Employing the professionals

Do you plan to go it alone (only really advisable if you're planning simply to redecorate or add new furnishings), or do you need the help of professionals—and, if so, which one is best for you?

◆ Architect: If you are remodeling the house and using an architect, she/he will oversee all the trades, including those dealing with the floor and wall finishes, and you should ensure this is included in the original agreement.

◆ Interior designer: An interior designer will work with you to find the right finishes for your needs, helping you to develop your style and taste. She/he will then help you to engage the contractors and oversee the work until it has been completed.

◆ **Builder:** If you are employing a builder, she/he may oversee the decorations and laying of floor finishes for you, or you may decide to go to the individual contractor yourself. The advantages of doing this are that you can negotiate the individual prices yourself, avoiding any builder markup, and you can be in charge of the work schedule. If you already have a relationship with the particular companies, this could mean you are less likely to be faced with no-shows. However, if you have a comprehensive building job on site, you may find it difficult to coordinate the individual contractors and you may also be less well versed than a professional builder at negotiating responsibilities between the contractors and the liabilities, should anything go wrong.

Smaller jobs

If you are planning simply to lay a floor or decorate a few rooms (or, indeed, the whole house) without taking on any building work, you will probably find it preferable to oversee the job yourself, engaging the professionals in the right order as you need them. The easiest way to manage this is to wait until each professional has finished their job before hiring the next. Even if the whole job takes longer, you may end up feeling less anxious.

Hiring professionals/contractors

The best way to vet out any professional is by personal recommendation. However, floor and tile suppliers can very often recommend professionals

ESTIMATED COSTS

◆ Whatever professionals you employ, you need comparable quotations.

◆ Start by writing as full a specification as possible. For example, if you are employing a decorator, you will need to state exactly which surfaces are to be painted in which finish (latex, eggshell, gloss), and with how many coats of paint. This will make the quotation process more reliable. It will also protect you further down the line once the job is on site, reducing any disputes where builder and client have a different understanding about what is and what is not included.

◆ If you are employing an architect or interior designer, they will write a detailed brief (specification) for you.

◆ If you are employing builders, they will organize all the other professionals, but, when bidding for services, you will need to write as full a brief as possible for the whole job.

◆ Give each builder/professional the same brief so you can compare one with another.

◆ When the quotations come in, read them carefully to check that they do indeed all include the same work. In reality, they are often a little different. Even given the same brief, some builders will cost all the elements, while others may omit some. Go through the specification and ask for more prices where they have been omitted to ensure you have properly comparable prices.

qualified to work with their materials.

◆ It is best to contact between three and six professionals for each job, and preferably six to ensure comparative quotes, as some drop out of the bidding for various reasons.

◆ Make sure they have proper certification, and that they have the relevant insurance.

◆ When you first contact them, let them know that you will want to see their bona fides and references before you engage them.

◆ Ask to meet them, and if they can take on the job and you feel you could communicate and work with them, then give them your brief/specification to review as well as your price estimate (see box on page 155).

◆ When the quotations are in and you have analyzed them, invite one or more to come and meet you again. The cheapest quotation is not always the best, nor is the most expensive.

◆ Ask yourself if you trust these people and would be able to work well with them through all the stresses and strains of building work; go and look at previous jobs; look at their workmanship, and ask the clients whether they had a good working relationship with them.

Payments

Before you engage a professional, ask what his/her terms are. Many will want an upfront payment for materials and then staggered payments until the job has been completed.

Never pay the whole amount in advance and always keep back the final payment until you have carefully inspected the job (with the job manager if you have one) and agreed that it meets your requirements.

BE READY FOR THE PROFESSIONALS

◆ For the smooth running of a job, make sure you're ready when the professionals arrive. Being ill-prepared will only cost you extra money, possibly result in damaged possessions, extend the job, and start you off on the wrong footing with your builders or contractor.

◆ Completely clearing the room to be decorated or given a new floor may be boring and inconvenient, but it is absolutely necessary for the smooth running of the job. Leaving "just the odd bit" to be moved the day work commences slows the job and irritates the workers. A clear room earns you plenty of brownie points and usually results in plenty of goodwill from the professionals as well as a smooth-running job.

PHOTOGRAPHY CREDITS

The publisher would like to thank the following photographers for supplying the pictures in this book:

Page 1 William P. Steele; **2** Simon Upton; **3** Anthony Cotsifas; **4 top** Michael Weschler; **4 bottom** Fernando Bengoechea; **5 top** Simon Upton; **5 bottom** Carlos Domenech; **6–7** Fernando Bengoechea; **8–9** Oberto Gili; **10** Gordon Beall; **10–11** Anthony Cotsifas; **12** Timothy Hursley; **13** Laura Resen; **14** Simon Upton; **15** Guy Bouchet; **16** Caroline Arber; **17** Jonn Coolidge; **18** Richard Felber; **18–19** Erica Lennard; **20** Tria Giovan; **21** Eric Roth; **22–23** Scott Frances; **24–25** Fernando Bengoechea; **25 top** Fernando Bengoechea; **25 bottom** Fernando Bengoechea; **26–27** Alec Hemer; **28** Firooz Zahedi; **29** William Waldron; **30** Tim Street-Porter; **30–31** Chuck Baker; **32 left** Jeff McNamara; **32 right** William P. Steele; **33** William P. Steele; **34** Dana Gallagher; **34–35** William Waldron; **36** Scott Frances; **36–37** Scott Frances; **37** Jeff McNamara; **38** William Waldron; **38–39** Scott Frances; **39** Tim Street-Porter; **40** Gordon Beall; **41** Carolyn Englefield; **42** Michael Luppino; **42–43** Jeff McNamara; **43** Nancy E. Hill; **44–45** Simon Upton; **45 top** Barbara and René Stoeltie; **45 bottom** Dana Gallagher; **46** Bob Hiemestra; **46–47** Anthony Cotsifas; **48–49** Dominique Vorillon; **50** Alan Weintraub; **50–51** Oberto Gili; **51** Jeremy Samuelson; **52** Dominique Vorillon; **53 top** Gordon Beall; **53 bottom** Eric Roth; **54** Eric Biasecki; **54–55** Michael Arnaud; **56** Oberto Gili; **56–57** Alan Weintraub; **57** Oberto Gili; **58** Scott Frances; **58–59** Simon Upton; **59** William Waldron; **60** Eric Boman; **60–61** Eric Piasecki; **61** Gordon Beall; **62** Eric Roth; **62–63** Scott Frances; **63 top left** Alan Weintraub; **63 top right** Alan Weintraub; **64** William P. Steele; **64–65** Gordon Beall; **65** Fernando Bengoechea; **66** Jeremy Samuelson; **67** Roger Davies; **68 left** Dominique Vorillon; **68 top right** Dominique Vorillon; **68 bottom right** Oberto Gili; **69** Antony Cotsifas; **70 left** Gordon Beall; **70 right** Luke White; **71** Christophe Dugied; **72** Gordon Beall; **73** William Waldron; **74** Christopher Drake; **74–75** Dominique Vorillon; **75 top** Michael Skott; **75 bottom** William Waldron; **76** Roger Davies; **77 top** Carlos Domenech; **77 bottom** Simon Upton; **78** Carlos Domenech; **78–79** William Waldron; **80** David Montgomery; **80–81** William Waldron; **82 left** Michael Luppino; **82 right** Jeff McNamara; **83** Christopher Drake; **84 left** Carlos Domenech; **84 right** Carlos Emilio; **85** Jacques Dirand; **86 left** Jacques Dirand; **86 right** Carlos Domenech; **87** Gordon Beall; **88** Simon McBride; **88–89** Robert Hiemestra; **89** Erica Lennard; **90** Gordon Beall; **90–91** Jonn Coolidge; **91** Andreas von Einsiedel; **92** Jonn Coolidge; **93 top** Buff Strickland; **93 bottom** Toshi Otsuki; **94** Jonn Coolidge; **95** Carlos Emilio; **96** Toshi Otsuki; **96–97** Robert Hiemestra; **97 top** Buff Strickland; **97 bottom** Toshi Otsuki; **98** Solvi dos Santos; **98–99** Robert Hiemestra; **99** Eric Boman; **100** Tim Street-Porter; **101** Anthony Cotsifas; **102–103** Simon Upton; **104 left** Caroline Arber; **104 right** Scott Frances; **105 top** Fernando Bengoechea; **105 bottom** Fritz von der Schulenburg; **106** Oberto Gili; **107 top** Christophe Dugied; **107 bottom** Christophe Dugied; **108** Steven Randazzo; **109** Christophe Dugied; **110 top** Antoine Bootz; **110 bottom** Scott Frances; **111** Laura Resen; **112 top** Christophe Dugied; **112 bottom** Chuck Baker; **113** Christopher Simon Sykes; **114** Laura Resen; **114–115** Barbara and René Stoeltie; **115** Simon Upton; **116** Lauren Resen; **116–117** Michael Weschler; **117 top** Christophe Dugied; **117 bottom** David Montgomery; **118** Simon Upton; **119 top** Thibault Jeanson; **119 bottom** Colleen Duffley; **120 left** Thibault Jeanson; **120–121** Laura Resen; **122** Thibault Jeanson; **122–123** Thibault Jeanson; **124** Christophe Dugied; **124–125** Simon Upton; **125 top** Thibault Jeanson; **125 bottom** Jacques Dirand; **126** William P. Steele; **127 top left** Dominique Vorillon; **127 top right** Dominique Vorillon; **127 bottom** William P. Steele; **128** Oberto Gili; **129 top** Jeff McNamara; **129 bottom** Carlos Domenech; **130** Oberto Gili; **130–131** Armstrong, Inc.; **132 top** Laura Resen; **132 bottom** Armstrong, Inc.; **133** Carlos Domenech; **134** Jonn Coolidge; **135** Scott Frances; **136** Scott Frances; **136–137** Jeff McNamara; **138** David Prince; **139 top** Eric Boman; **139 bottom** Jeff McNamara; **140** Tara Sgroi; **140–141** Oberto Gili; **142** Eric Boman; **143** Gordon Beall; **144** William Waldron; **144–145** Alex Hemer; **146** Eric Boman; **146–147** Jeff McNamara; **147** Christina Schmidhofer; **148** William Waldron; **148–149** Jeff McNamara; **149** William Waldron; **150** Erica Lennard; **151** Oberto Gili; **152–153** Laura Resen; **156** Tria Giovan.

INDEX